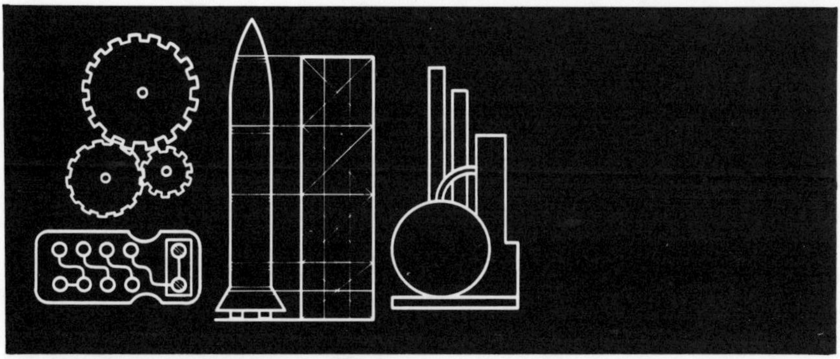

CAREERS IN ENGINEERING AND ENGINEERING TECHNOLOGY

CAREERS IN ENGINEERING AND ENGINEERING TECHNOLOGY

by Mary McHugh

FRANKLIN WATTS | NEW YORK | LONDON | 1978

Photographs courtesy of:

Bell Laboratories: pp. 4, 9, 14, 21, 24, 39; New Jersey Institute of Technology: p. 11; Mary McHugh: pp. 17, 18, 27, 47; New York Institute of Technology: pp. 36, 42, 43, 44, (Edward J. Edahl) 49.

Library of Congress Cataloging in Publication Data

McHugh, Mary.

Careers in engineering and engineering technology.

(A Career concise guide)

Includes index.

SUMMARY: Discusses various jobs in the engineering field, the training needed, and starting salaries. Includes a list of schools providing training.

1. Engineering—Vocational guidance—Juvenile literature. [1. Engineering — Vocational guidance. 2. Vocational guidance] I. Title.

TA157.M25 620'.0023 77-16254

ISBN 0-531-01424-X

Printed in the United States of America
6 5 4 3 2 1

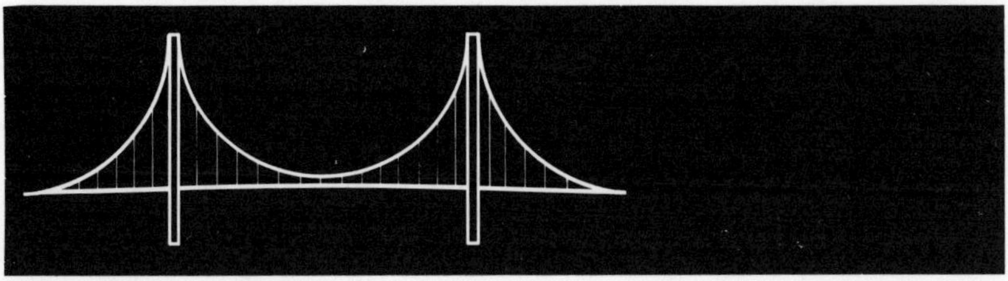

Contents

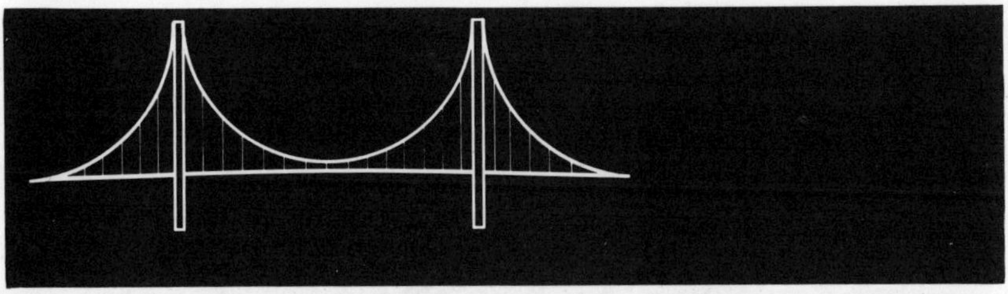

Careers in Engineering

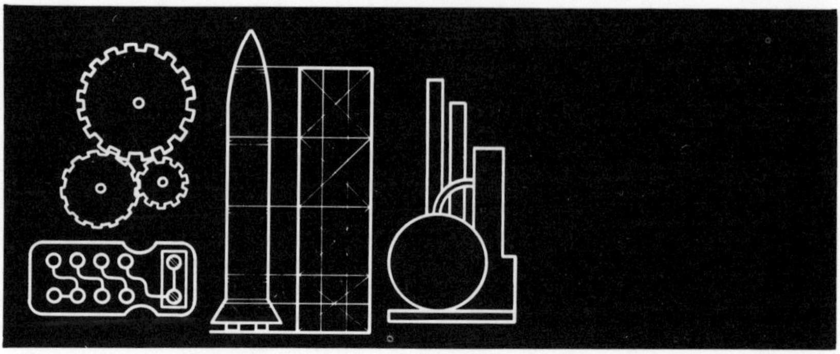

Engineers make dreams come true. All the comforts, necessities, luxuries, and conveniences of modern life were made possible by an engineer turning a scientific theory into a reality. Without engineers, there would be no roads, no Corvettes, no pocket calculators, no stereos, no television, no telephones, no rockets to the moon, no movies, and no Burger Kings.

Literally every manufacturer, power plant, and communications company depends on engineers to produce new products, improve old ones, and perfect systems for conveying power and sound, and in 1977 there were more jobs available to engineering graduates than in any other field.

As in most careers, certain kinds of people are more likely to become engineers than others, and it is often clear during childhood whether or not a person is heading in that direction. Certain children—usually boys, since we are just beginning to encourage little girls to explore the world as fully as their brothers—show a great deal of curiosity about how and why things work. You know if you're one of those unusually curious people, because you probably drove your mother crazy taking clocks and radios apart when you were a child.

Not all engineers started life pulling things apart, but most did show an aptitude for math and science and took as many courses in these subjects as were offered in high school. Some went to school on Saturday to take extra classes. Many future engineers found after-school and summer jobs in TV and stereo shops or at gas stations. Motors and switches held a great deal of fascination for these students and they took them apart and put them back together the way some people do jigsaw puzzles.

Steve Pope, an electrical engineer who graduated from Cornell University in 1977, always knew he wanted to be an engineer. "I've always loved physics and geometry and all sorts of science and math," Steve says. "It wasn't that I thought, 'Maybe I'll give engineering a try.' I knew I wanted to do it from the start."

Steve advises high school students to write to professional engineering societies to ask for information about the field each society specializes in. These organizations accept students as members and publish newsletters to keep them apprised of scholarships and contests.

You might also write to the Junior Engineering Technical Society at 345 East Forty-seventh Street, New York, New York 10017, to ask for their monthly newsletter for high school students.

There are over 230 engineering programs in the United States. (They are listed in the appendix of this book.) Unlike many other professions, engineering is taught in undergraduate programs, usually requiring four years to complete. You can earn a Bachelor of Science degree in one of the many engineering specialties.

In most four-year programs, the courses during the first two years are the same for all disciplines: math, physics, chemistry, the humanities, English, and social sciences. The curriculum for the last two years concentrates on the area of specialization. (A sampling of these courses is included in each chapter.)

There are also five-year programs, some of which award both a Bachelor of Science and a Master of Science degree.

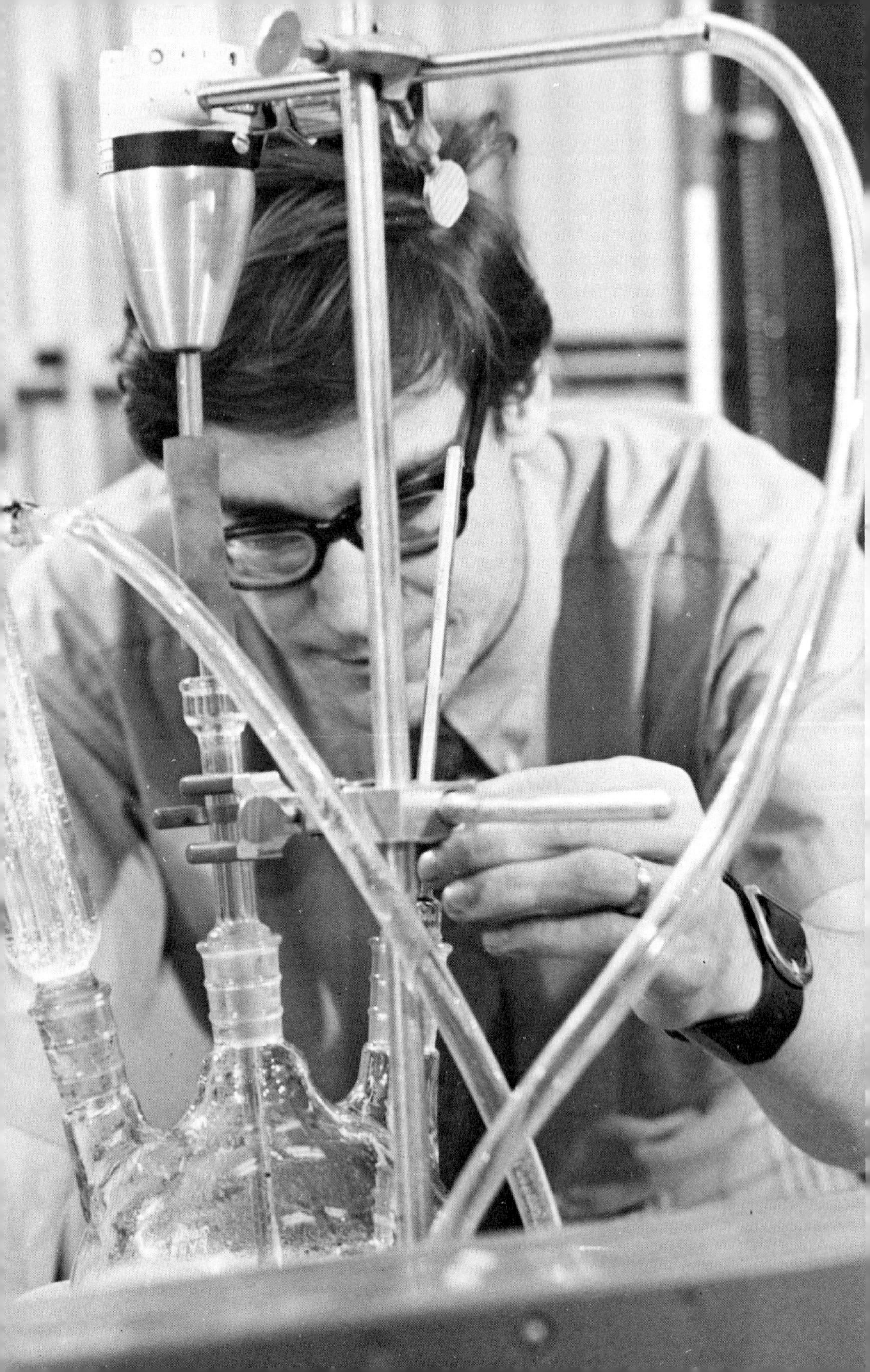

Other programs include a three-year liberal arts curriculum and a two-year engineering program, in which the student receives a Bachelor of Arts and a Bachelor of Science degree. Work-study cooperative programs also take five or six years.

Scholarships and other financial aid are available, especially to minority students, as a great effort is being made by the universities and by industry to encourage more minority students to become engineers. To find out about financial aid, write to the National Society of Professional Engineers, 2029 K Street, NW, Washington, D.C. 20006, or to your own state Society of Professional Engineers. There is also an excellent pamphlet available for fifty cents called "Need a Lift," published by the American Legion, Department S, P. O. Box 1055, Indianapolis, Indiana, which gives information on state and federal loans, scholarships, and other financial aid available.

Some engineers have found that a master's degree, either in their field or in business administration, is a valuable degree to have. About 20 percent of engineers go on to get an advanced degree, often at their company's expense. Doctorates in engineering are usually required for research or teaching. Salaries are higher for those with a master's, and much higher for those with doctor's degrees in engineering.

Starting salaries for engineers these days are excellent. In 1976, graduates received between $13,000 and $14,000 to start, and in 1977 graduates were being offered $14,000 to $15,000. Each specialty brings a slightly different starting salary, according to the demand at the moment. In 1977, chemical engineers were highly valued because of our increasing national concern with pollution control, and aerospace engineers are not as much in demand as they had been in the past because of cutbacks in space and defense funds.

In recent years, the highest salaries in many engineering specialties have been going to women. The median salary for

An engineer at Bell Laboratories working on the recycling of plastic.

women engineers in 1976 was $14,210—2 percent higher than for men. Because industry and universities must hire more women or lose government contracts, the few women to graduate with engineering degrees are in great demand, and companies are willing to pay more to get them. Only 1 percent of all engineers today are women, but the percentage of women students in engineering programs has increased to 20 percent so they won't be quite so scarce in the future.

Is engineering a good field for women? Why are there so few women in it? If you were to ask an old-line engineer, he would tell you that women just aren't smart enough and that they are especially bad in math. And that even if they *were* good, he wouldn't hire them, because he wouldn't be able to swear on the job. Ask any woman engineer why there are so few of her kind, and she'll smile wryly and tell you that she met with discouragement every step of the way—from the time she innocently mentioned the subject to her parents and was told that teaching was more suitable for a girl, until she was graduated from engineering school and met with the stubborn resistance of male engineers who simply didn't want a woman around.

All that is changing, however, because the law has forced employers to hire more women, and the competence and intelligence of women pursuing engineering careers have convinced most old die-hards of women's value in the field. For more information about opportunities for women write to the Society of Women Engineers, 345 East Forty-seventh Street, New York, New York 10017.

You will often see the letters "P.E." after the names of some engineers. These stand for Professional Engineer and can be earned after four years of experience and a two-day, sixteen-hour examination. You can take the first part of the exam at the end of your senior year in college if you like, and the second part two to four years later (depending on the state law), or you can take the whole exam later in your career. A license as a Professional Engineer is necessary if your job affects the safety or health of the public. Many people equate a P.E. license with a master's degree.

Since there isn't room to discuss every specialty in engineering, this book will describe the eleven major fields. This should help you narrow your choice if you are considering going to engineering school but haven't yet picked your area of specialization. The following pages will discuss the kinds of jobs available, the training needed, and the starting salaries being offered in the fields of electrical, chemical, mechanical, civil, industrial, aerospace, ceramic, agricultural, metallurgical, mining, and biomedical engineering. The second part of the book will be concerned with careers in engineering technology, and will give detailed information on electrical, chemical, mechanical, civil, air conditioning, heating and refrigeration, instrumentation, and aerospace technology.

For general information about engineering careers, write to:

National Society of Professional Engineers
2029 K Street, NW
Washington, D.C. 20006

Engineers' Council for Professional Development
345 East Forty-seventh Street
New York, New York 10017

CAREERS IN ELECTRICAL ENGINEERING

Electrical engineering is a field with many exciting possibilities for the future. It is the largest branch of the profession, with over 230,000 engineers, and holds many attractions for those considering this specialty, including the major advantage of a job offer after graduation.

Large companies such as utilities, telephone companies, manufacturers of electrical appliances and equipment, radio and TV companies, as well as research institutions, universities and government agencies are looking for talented graduates of electrical engineering programs. There are a wide variety of jobs for these graduates in the design, development, supervision, and manufacture of new and fascinating products such as smoke detectors, electric systems on satellites, submarine propulsion systems, mass-transit trains, radarwave ovens. Starting salaries for those with bachelors degrees range from $14,000 to $15,500, depending on the specialty and amount of training. About 40 percent of electrical engineers have an advanced degree, especially if they are in research, teaching, or management.

Computers have revolutionized almost every industry, and a good specialty for the electrical engineer these days is computer architecture and design. Steve Pope chose this specialty because he knew it was a growing field. According to Steve: "It takes a lot of specialized study, but you can work for any company that uses or makes computers. You can either design the computers themselves—called hardware by engineers—or you can design the programming, or software. Most computer engineers work for a firm like IBM which designs and manufactures computers to sell to other companies."

Miniaturized circuitry is another brand-new field which involves the design, development and production of devices such as pocket calculators, guided-missile systems and electronic TV games. Glen Dash, a twenty-four-year-old electrical engineer, developed an economical version of an electronic TV tennis game for a toy manufacturer when he was still a senior at the Massachusetts Institute of Technology (M.I.T.). He is now chief engineer for the company and spends his days thinking up new and ingenious electronic games and toys.

In the field of communications, the electrical engineer has always been needed to design the intricate systems of wire, both aboveground and underground; undersea cables, radios, and satellites which transmit sound and pictures across continents; and switching systems which carry the voice from one place to another without distortion.

By the beginning of the twenty-first century, electricity will play an even greater role in our daily lives. The same electronic controls that space scientists used to maneuver the shovel that dug up samples of soil on Mars could conceivably roll a table into the room with your dinner on it.

Power companies, too, depend on electrical engineers to

Victor Ransom, an electrical engineer at Bell Laboratories, explains an internal measuring system for the company's traffic department.

help design the systems that carry hydroelectric, steam, electric, or nuclear power to industries and homes and to supervise the installation of these systems.

If this field appeals to you, you should take as much math as you can in high school—algebra, plane and solid geometry, trigonometry, and calculus. You should also take physics, chemistry, and English. Check with your guidance counselor for other courses that might be required, such as a foreign language.

In college, besides the general courses common to all engineering students such as English, social sciences, math, physics, biology, chemistry, computer science, economics, political science, and history, you would take more specialized courses later on. These would include applied math—such as differential and integral calculus, statistics and probability —as well as electromechanics, electromagnetic theory, digital computing, and circuit theory.

For some reason, women don't seem to choose electrical engineering as readily as men do. However, as engineering itself becomes more accepted as a field for women, there will probably be more who will choose this specialty.

For more information, write:

Institute of Electrical and
Electronics Engineers, Inc.
345 East Forty-seventh Street
New York, New York 10017.

CHEMICAL ENGINEERING

In 1977, starting salaries for chemical engineers were $16,500, higher than for any other specialty except petroleum engineers. There are now about seventy thousand chemical engineers, two-thirds of whom are in management, earning top salaries throughout their careers.

Why are chemical engineers so valuable to industry? Probably because they work with other types of engineers to solve the world's most serious problems. Chemical engineers play a large part in pollution control, sewage disposal, the recy-

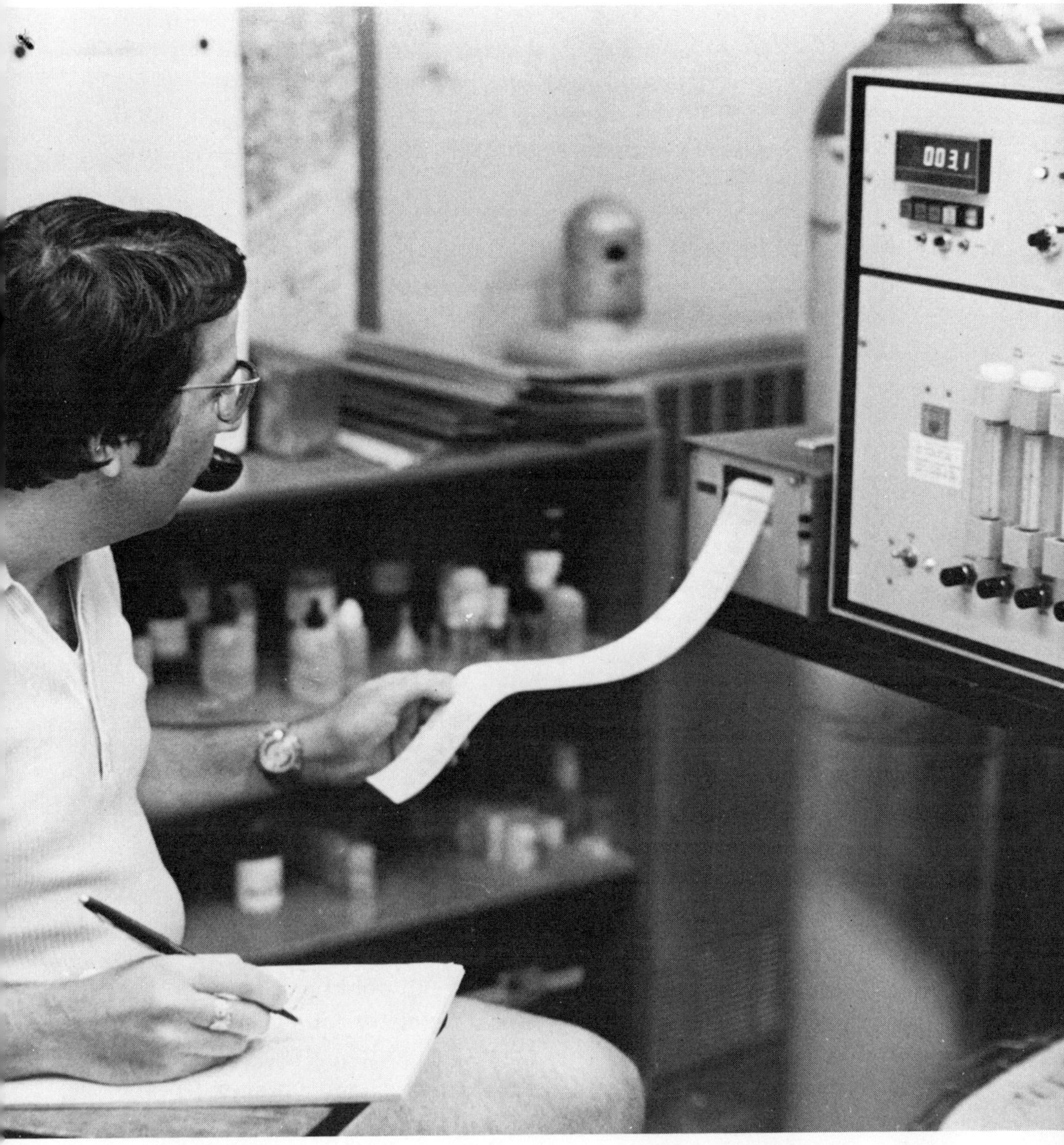

Professor Richard Trattner, associate professor of chemistry and environmental science at New Jersey Institute of Technology, studies results of water samples taken at the Meadowlands, N.J.

cling of waste materials, increasing food production with the use of improved fertilizers, the replacement of dwindling raw materials with plastics and other synthetics, the creation of new energy sources, desalinization of seawater for irrigation purposes, and development of antibiotics and other lifesaving drugs.

All kinds of companies employ chemical engineers—mainly chemical and petroleum companies, but also companies connected with the aerospace industry, as well as atomic energy plants, steel mills, food companies, textile, paper and pharmaceutical firms, hospitals, laboratories, and banks.

Lawrence Donnelly, who earned a master's degree in chemical engineering at Yale, heads his own consulting firm. He started out working on defense projects, including chemical rockets, but when it became apparent that the government was cutting back on defense, he wisely switched over to the civil sector. "I decided to get out of defense work, and was put in charge of a sewage-treatment project. We designed and patented a shipboard sewage-treatment system that was less bulky than ordinary municipal sewage plants.

"Then I started my own business as a consulting engineer working in the field of industrial-waste treatment. We advise companies on ways to meet federal and state regulations regarding the chemical treatment of sewage discharged from plants."

The usual preparation in high school for a chemical engineering program in college is two years of algebra, one year of geometry, half a year of trigonometry, one year of physics, and one year of chemistry.

Some of the courses included in chemical engineering programs in college are organic chemistry, analysis of separation processes, thermodynamics, chemical process evaluation, reaction kinetics and reactor design, and chemical process synthesis.

To find out more about this field, write to:

American Institute of Chemical Engineers
345 East Forty-seventh Street
New York, New York 10017

MECHANICAL ENGINEERING

Where can mechanical engineers be found? Everywhere. Second only to electrical engineers in number, there are now about 210,000 mechanical engineers in the United States, and the Department of Labor estimates that the country will need 280,000 by 1985. The possibilities are almost unlimited because mechanical engineers can work for automotive, power, telephone, aerospace, mass transit, textile, computer, office machinery, cosmetics, and construction companies.

Your job might be helping to design a monorail system to streamline urban transportation. You might help design a two-man diving sphere to explore the ocean for new sources of energy. Or, you might help develop a system of machinery that will convert the energy of water, coal, oil, gas, nuclear reactors, and the sun into power for heating or cooling our homes, for providing a cheap fuel for our cars, or for lighting up our cities.

Besides working on the design and development of new machines and the improvement of old ones, mechanical engineers also find positions as plant managers or plant superintendents, supervising the production of machinery. Or, as in other engineering disciplines, with an education in mechanical engineering you might become a salesperson, having the technical knowledge necessary to explain highly complex machinery to potential customers.

Then again, you might end up as chairman of the board or president of General Electric. Why not? Many mechanical engineers, as well as engineers in other specialties, eventually find themselves with executive positions. Charles Larson, executive director of Industrial Research Institute, is a good example of an engineer who became intrigued with the challenge of management. Trained as a mechanical engineer at Purdue University, he soon realized that he wanted to use his knowledge of engineering in an executive capacity—to advise others rather than doing the actual research himself. He says:

> I became interested in business and economics and decided to get a master's degree in business administration.

Business school was certainly different from engineering school. It involved a lot of talking, a lot of reasoning, and a lot of plain common sense.

Engineers can progress from a project engineer or group leader, to chief engineer to vice-president of engineering, to executive vice-president of the company, and then to president and chairman of the board. But even when you become an executive, people still look on you as an engineer. You're there because you are an engineer, and that's what the company needed, rather than a legal or financial man.

To prepare for a career in mechanical engineering, you would take the basic math, physics and chemistry courses in college. You would also study solid mechanics, fluid mechanics, thermodynamics, heat transfer, systems and controls, materials, electricity magnetism, and design; plus English, graphics, computer languages, and courses in the humanities.

In high school you should take as much math as possible, of course, including algebra, plane and solid geometry, trigonometry, and calculus. You should also take physics, chemistry, English, and whatever other science courses interest you.

Starting salaries for mechanical engineers with a bachelor's degree range from $12,000 to $15,000. Salaries for those with a master's start at $16,000 or $17,000, and Ph.D.'s are hired at $20,000 and up.

For more information, write:

American Society of Mechanical Engineers
345 East Forty-seventh Street
New York, New York 10017

John Formica (right), who got his B.S. in mechanical engineering and his M.E. in Industrial Engineering, works with Jesse Russell, an electrical engineer at Bell Laboratories.

CIVIL ENGINEERING

"Civil engineering has the greatest impact on people's lives because we build roads, plan water and sewer projects, and construct housing needed by the public," says Rick Alberts, a young civil engineer in Florida, with a special interest in the impact of building projects on the environment. He works for a company that advises major developers and builders on the best possible way to construct a road, housing project, or airport so that it will have the least adverse effect on the environment and the local community.

Civil engineers are responsible for the design and construction of our country's bridges, dams, highways, buildings, railroads, tunnels, airports, and docks. They design irrigation projects and systems for waste disposal. They design, construct, and maintain power plants.

The complicated problems of modern cities require new and resourceful solutions. Roosevelt Island, once Welfare Island, off Manhattan's East Side, is an example of the inventive thinking of engineers. To minimize pollution, automobiles are banned from the island and electric buses are used to transport residents. An aerial tramway provides a particularly innovative means of taking people across the river to Manhattan. The problem of waste disposal was solved by a system of pipes that deposit all refuse in one building. There are no garbage trucks and no smelly piles of garbage awaiting pickup.

Civil engineers must solve incredibly complicated problems, such as carving a tunnel through the Rocky Mountains, or building a highway in Alaska where most of the soil is frozen all year long. They construct rigs that can search for oil miles below the ocean's surface. They design and supervise the construction of skyscrapers, like Chicago's Sears Tower, which is the highest structure ever built.

The challenges faced by civil engineers make this a fascinating field. Carl Whitehead, vice-president of Ebasco Services, a power company in New York, stresses the need for more civil engineers, particularly power-plant engineers. "This is a good field to go into," he says. "The forecast is for a

Frank Blesso, a civil engineer, who is Deputy Redevelopment Director of the Department of Community Development in Paterson, goes over the plans for an urban redevelopment project.

A civil engineer and a landscape architect work together on urban redevelopment.

shortage of civil engineers in 1982, especially power-plant engineers." The Department of Labor estimates there will be a need for 235,000 civil engineers in 1985. There are about 180,000 now.

The starting salary for civil engineers in 1977 was about $14,000 and was higher for those with a master's or Ph.D. degree.

The curriculum in most civil engineering programs includes math, chemistry, physics, history, economics, government, English, sociology, psychology, basic engineering theory, and analysis and design. More than one-third of all civil engineering graduates go on to get a master's, either in their field or in business administration to prepare for management positions, where salaries of top executives are over $35,000 a year.

Many civil engineers are required to be licensed as Professional Engineers because their work so often involves the health and safety of the public.

For more information, write:

American Society of Civil Engineers
345 East Forty-seventh Street
New York, New York 10017

INDUSTRIAL ENGINEERING

Industrial engineers were once called "efficiency experts," because it was their primary job to provide the highest level of productivity at the lowest possible cost, without sacrificing quality or safety.

Today, the responsibilities of industrial engineers have broadened to include other aspects of running a successful business. Besides planning the layout of a production plant, and working out incentive plans, they determine the specific tools, machinery and supporting facilities to be used. When they develop the plant layout, they make sure the numerous systems involving production, materials handling, electrical power, air conditioning, and heating can be easily maintained and serviced. They develop alternate ways to mechanize or automate as many production processes as possible to improve plant output capability.

Industrial engineers often develop the controls to assure the quality of the end product during the intermediate steps of manufacturing, assembly, labeling, and packaging, as well as selecting the proper machines and systems to perform the production processes. They provide top management with the background information needed to decide whether to expand one plant or five. They can advise the president of a company how much an improvement in facilities will save in operating costs.

Engineering colleges today offer specific training in industrial engineering, although there are a large number of practicing industrial engineers who first trained as electrical, mechanical, chemical, or civil engineers.

One industrial engineer who had trained as a mechanical engineer had a unique challenge. While working for a direct mail order house he had to assemble a mechanical system of equipment—for a special nationwide promotion mailing for *Reader's Digest*—which would place two bright new pennies in with each piece of mail, showing through the window on the face of the envelope. Fifty million envelopes were mailed.

"The object was not to invent a new machine or process," he said, "but to assemble a production system from slightly modified equipment components that were already manufactured. That's the essence of industrial engineering."

In addition to basic engineering courses, industrial engineering programs include courses in applied mechanics, fluid mechanics, thermodynamics, materials engineering, electronics, facilities design, production control, scheduling theory, engineering economics, inventory systems, materials processing, and operations research.

To find out more about this challenging and rewarding specialty, write to:

American Institute of Industrial Engineers
25 Technology Park
Atlanta Norcross, Georgia 30071

Frank Jana, an industrial hygiene engineer at Western Electric, checks the sound level in a wire mill.

AEROSPACE ENGINEERING

Remember when "Fly Me to the Moon" was just a song? Now, thanks to aerospace engineers, we've been to the moon, and spaceships have explored the planet Mars. There are communications satellites, weather satellites, and an earth-resource satellite which surveys our planet's water resources, energy sources, forests, deserts, and so on. The growth of technology in this field in the last ten years has been extremely rapid, and in spite of the cutbacks in defense funds, the aerospace industry is thriving in other areas.

Aerospace engineers are also involved in the design and manufacture of crafts that explore the ocean's depths, and of high-speed trains. As well, they are at work on the development of vertical and short -takeoff and -landing aircraft and quiet short-haul air-transportation systems. Engineers work on aircraft and spacecraft from the design and planning stages to the final assembling and testing.

Don Alberts, an aerospace engineer with General Electric, encourages young men and women to consider this specialty. "When there are regular shuttle flights, you'll see a drastic change in attitude toward aerospace," he says. "It will be more like commercial aviation—less glamorous and economically sounder."

Don got his degree in civil engineering and he feels it is best to earn a bachelor's degree in civil, mechanical, structural, or electrical engineering to provide a broader base of knowledge, and then go on to specialize in aerospace by working for one of the large airplane manufacturers like Lockheed or Boeing. Many aerospace engineers go on to get advanced degrees in their field because the really exciting work is in new development.

To prepare for a career in aerospace engineering, you should take the usual math and science courses in high school—algebra, geometry, trigonometry, physics, chemistry, and biology—as well as English and history.

In a college aerospace engineering program, some typical courses for your junior and senior year would include fluid mechanics, heat transfer, electrical circuits, aeronautical laboratory, gas dynamics, electronics, modern physics, aerospace propulsion systems, boundary layer theory, and introduction to astronautics.

Starting salaries for aerospace engineers are around $13,000 or $14,000 a year.

If you want more information, write to:

American Institute of Aeronautics and Astronautics
1290 Avenue of the Americas
New York, New York 10019

CERAMIC ENGINEERING

Wanted: an automobile engine that will get fifty miles to the gallon. All kinds of engineers are concerned with solving this energy problem and reducing pollution, but ceramic engineers are in the forefront of the battle, because they are working with materials that are economical, readily available, and are extremely strong and temperature resistant. Ceramics are beginning to replace plastics as an important manufacturing material. This is because of the shortage of petroleum from which plastics are synthesized.

A project that concerns us all is the perfecting of a turbine engine with ceramic parts that can withstand temperatures up to 3,000 degrees Fahrenheit (1,649 Celsius). When ceramic engineers develop a way to build such engines economically, automobiles will get fifty miles to a gallon of fuel—but not necessarily gasoline, because scientists are searching for more plentiful substances. The fuel will burn without pollutants at that high temperature, eliminating the need for emission control in cars.

In the struggle to find more efficient and economical means of transportation, ceramics are being used in spark plugs, jet turbines, and nuclear propulsion systems.

Companies throughout the country need ceramic engineers to work on the design, development and improvement of such modern inventions as self-cleaning ovens, fiber-glass tires, prosthetic devices, nose cones for rockets, and solar heat collectors. The U.S. Department of Labor estimates that

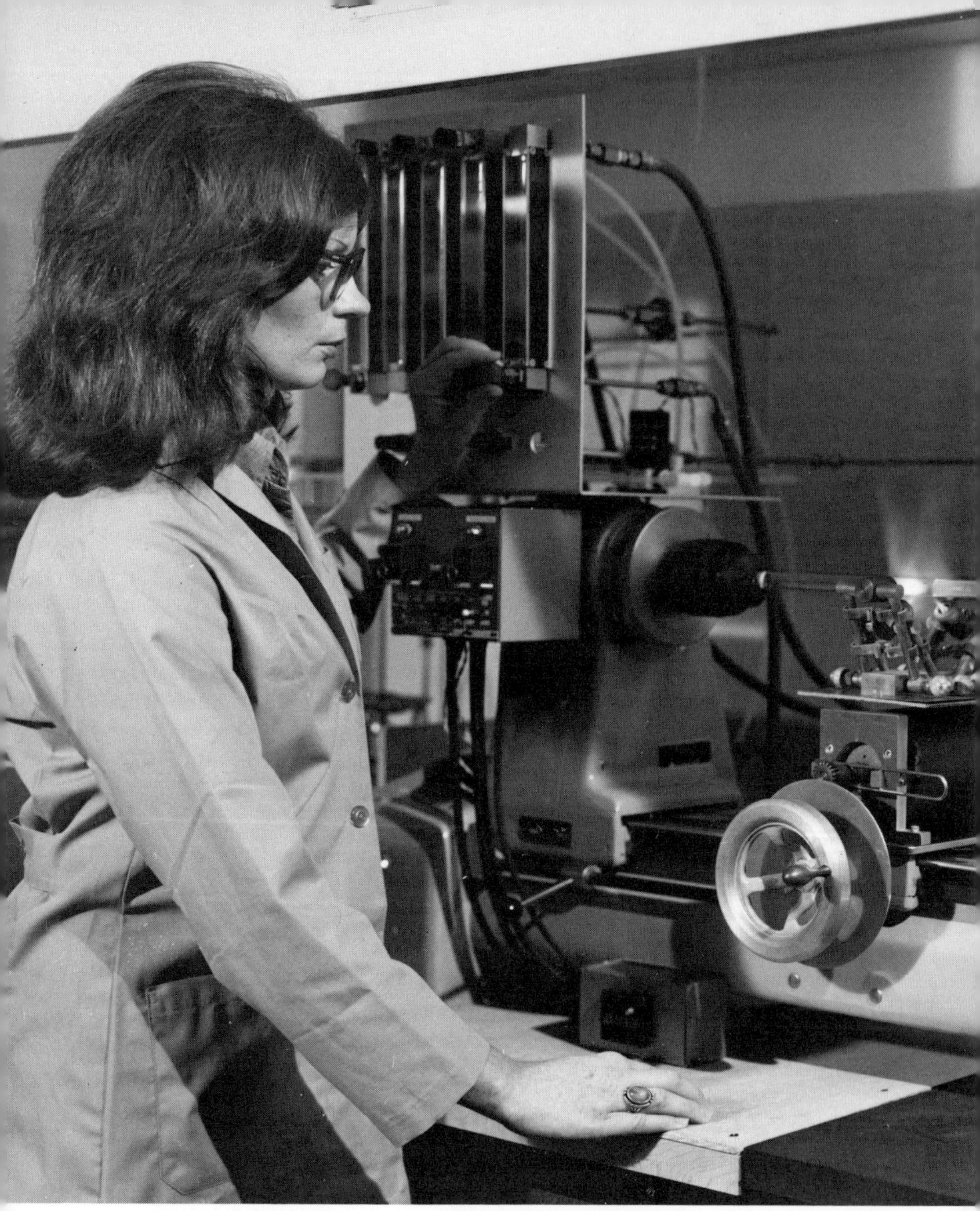

Suzanne Nagel, a ceramic engineer at Bell Laboratories, Murray Hill, N.J., doing research on optical fibers.

there will be a need for fourteen thousand ceramic engineers in 1985. There are about twelve thousand now, and at present the engineering schools are not turning out enough of them to fill the four hundred positions that will open each year between now and 1985. Because it is such a new field, it is often overlooked by engineering students.

Beginning salaries for ceramic engineers reflect the need for specialists in this field. The starting salary for ceramic engineers in 1976 was second to that of chemical engineers.

Dr. Suzanne Nagel is a ceramic engineer with Bell Laboratories at Murray Hill, New Jersey. She earned a Ph.D. because she knew she wanted to do research. She is presently working with optical wave-guides—thin glass filaments that can transmit many times the number of telephone messages carried by wires and cables, a vitally important development in our overcrowded, urbanized society. She says, "I enjoy working in research. It's the closest thing to being in a university. Here I have the freedom to design my own research."

Dr. Nagel's five-year engineering program at Rutgers, which earned her a B.A. as well as a B.S. degree, included basic engineering courses, with special courses in glass, refractories, whiteware, and porcelains, plus a liberal arts curriculum. In graduate school at the University of Illinois, she earned her doctorate studying the crystallization of glass.

To find out more about this specialty, write to:

National Institute of Ceramic Engineers
65 Ceramic Drive
Columbus, Ohio 43214

AGRICULTURAL ENGINEERING

Our world is running out of food and natural resources. Though the United States is approaching zero population growth, the world population may reach six or seven billion by the year 2000, more than twice the number now in existence. The challenge of finding new ways to produce food and clothes for all those people while reducing pollution and protecting the environment is the job of the agricultural engineer.

Someday there may be climate-controlled plastic domes enclosing special crops that are difficult to grow. Huge, driverless machines controlled by a computer will harvest a farmer's crops more efficiently and economically than they are today. Agricultural engineers will struggle with the problems of soil and water conservation, reuse of land, construction of temperature-controlled buildings for storage and shelter, more efficient methods of food processing, and power and machinery design for increasingly automated farm functions.

Agricultural engineering is a relatively new specialty. Although engineers have worked with farmers for years to help them solve their problems, they were usually trained as civil or mechanical engineers and learned the special problems of the farmer from experience. Now, colleges and universities offer degrees in agricultural engineering.

Besides the basic engineering courses in the first two years, the curriculum for the junior and senior years includes the study of biology, plant or animal physiology, soils, control of building environment for plants and animals, soil and water conservation, building design, machinery design, power-unit design, food processing, and systems engineering.

A typical project for students in an agricultural engineering program, for instance, might be the design of a hand-operated pump to be used in an underdeveloped country like Bangladesh, or a highly sophisticated, mechanical blueberry sorter. Professor William J. Roberts, chairman of the Biologi-

Joel Simpkins (left), an agricultural engineer, shows his solar energy project to Professor William J. Roberts, Chairman of the Biological and Agricultural Engineering Department at Cook College, Rutgers University, in New Jersey.

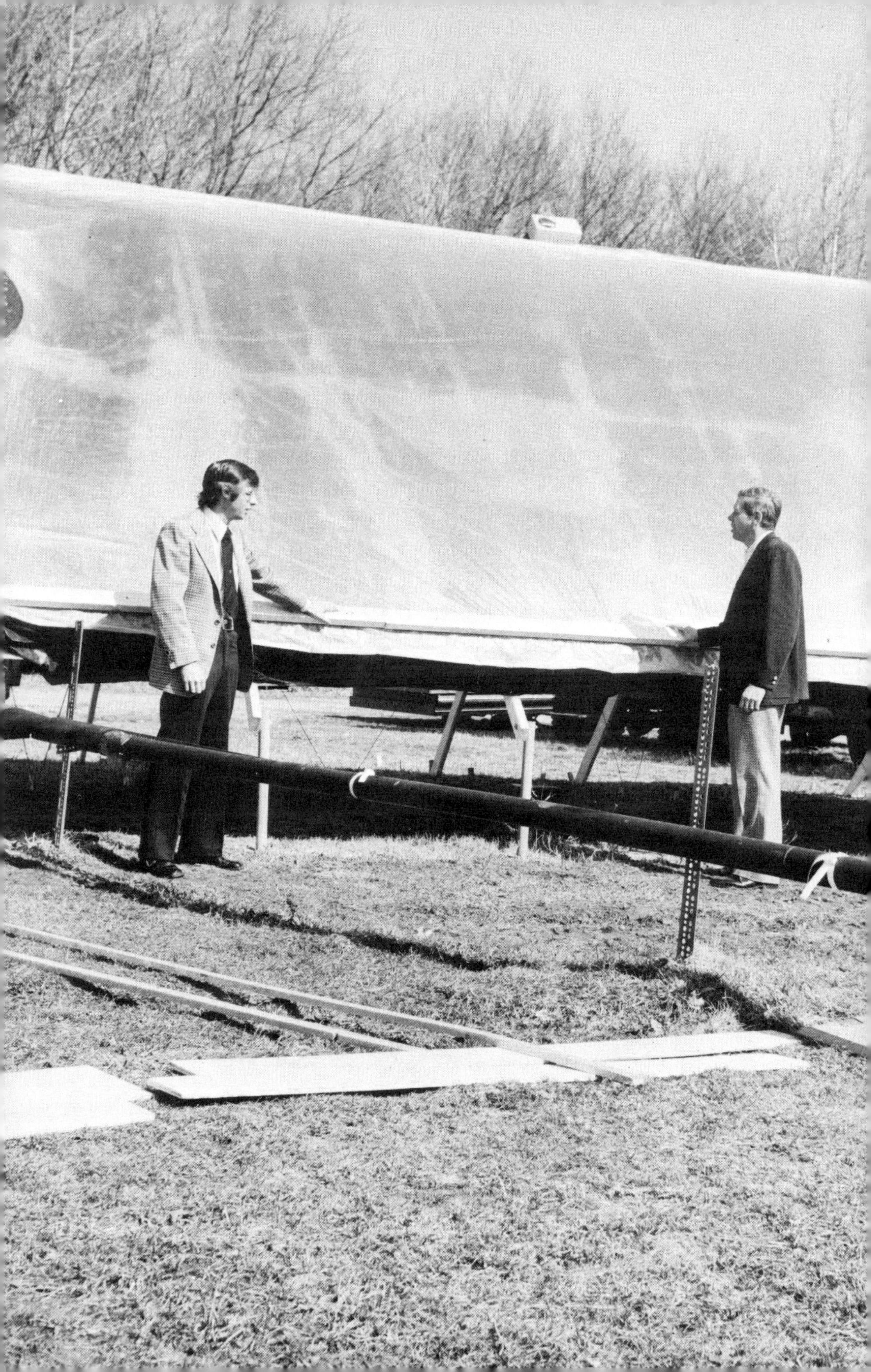

cal and Agricultural Engineering Department at Cook College, Rutgers University, in New Jersey, points out the growing need for people in this specialty: "As the world's food production becomes centralized in certain areas, storage becomes more and more important. Grain, for instance, must be stored at a particular moisture content. A civil engineer could design the building, but the agricultural engineer can design the building and also plan the proper environment to maintain the grain."

Joel Simpkins has a master's degree in agricultural engineering, and is working on a solar energy project at Rutgers. Joel says: "We're trying to reduce the amount of fossil fuel needed to heat a greenhouse by supplementing it with solar energy. We have to do it economically though—that's the engineer's job."

Although a farm background is not a requirement for an agricultural engineer, Joel worked on farms all his teen-age years. He recommends courses in math, physics, chemistry, and biology in high school to prepare for agricultural engineering. Involvement in high school clubs such as Future Farmers of America and 4-H clubs is also a good idea.

Beginning salaries for agricultural engineers are comparable to those of other engineers, around $13,000 to $14,000 a year. Large manufacturers of farm equipment such as International Harvester and Ford are looking for agricultural engineers, as are food-processing companies like General Foods, power companies, colleges and government agencies like the Agricultural Research Service, the Soil Conservation Service, and the Environmental Protection Agency.

"This is an interesting field," Professor Roberts says. "Requests come across my desk every day for graduates of our program to do research in livestock management and housing, environmental control, solar energy, soil conservation, irrigation, and food production."

For more information, write:

American Society of Agricultural Engineers
2950 Niles Road
Saint Joseph, Michigan 49085

METALLURGICAL ENGINEERING

"I think metallurgical engineering is a superb field," says Paul Goldberg, a consulting metallurgical engineer. "Metals constitute ninety percent of all known elements. Common things all around us, from automobiles to kitchen utensils, to aircraft and workshop tools, are made of metal. I'm somewhat of an empiricist—I like to hold the things I work with, rather than working only with abstract concepts."

Paul is often asked to testify in court as an expert witness in cases involving accidents in which failure of a metal part has occurred. He also uses his training as a metallurgical engineer to advise corporations on solving any problems that involve metal. He might analyze the cause of corrosion, for instance, and recommend a method of stopping it.

The study of metallurgical engineering includes learning how to extract metals from ores, refine them, form alloys, cast, fabricate and heat-treat metals, as well as design the plants that carry out these processes.

There is a growing tendency in colleges to include metallurgical engineering in the materials-science department, where students are also taught the properties and uses of ceramics and polymers (plastics, silicons, etc.).

Because of the shortage of natural resources, there is a real need for people trained to discover ways of conserving metals and minerals. The ingenuity of engineers is needed to find ways to extract metals from low-grade sources while still protecting the environment.

Because of new state and federal regulations regarding air and water pollution, metallurgical engineers are hired to monitor the gases emitted by heavy industry and sample them to make sure they meet the requirements of environmental-protection agencies. The engineers also test the water that is discharged from the plant, and treat the air to remove pollutants, as well as precipitating soluble metals out of the water.

Metallurgical engineers are hired at starting salaries of $13,000 to $14,000 by companies like Phelps Dodge, Anaconda, Kennecott, American Metals, Gulf Resources, U.S. Steel, Alcoa, IBM, and RCA. They might also work for the federal government in army and navy laboratories. There are now about ten thousand engineers in this field.

For more information, write:

American Institute of Mining,
Metallurgical, and Petroleum Engineers
345 East Forty-seventh Street
New York, New York 10017

MINING ENGINEERING

Sandra Gabbet, a geological engineer, enjoys her job with the French American Metal Corporation. She is part of a team involved in uranium exploration. "That entails interpreting oil logs and going into the field to 'sit wells'—that's when you go out to the rigs and supervise the drilling hands and the geophysicists who are doing the electrical analysis. You also check samples as they come out and identify the kind of rocks they're drilling in. We look for uranium in crystalline rocks or sedentary basins."

How is the job situation for women mining engineers these days? "Very good," Sandra says. "At least in the large companies which are required by law to hire a certain number of women in order to make up for past discrimination. Oil companies especially are looking for women to fill management positions. I was interviewed by five companies and got four job offers. A lot of companies want graduates with master's degrees and I plan to go back to school for mine after I've worked for a year. The smaller companies are not so anxious to hire women because they have reservations about sending women out into remote areas. And there's still a strong taboo against women going into the mines."

Mining engineers analyze the stability of a rock slope and recommend the best place for a mine or well. A mining engineer plans everything from the actual installation to the roads and houses that will be built for the miners moving into the area.

Because of the energy shortage, mining engineers are needed to discover new ways of extracting coal and uranium, metals, gypsum and limestone, phosphates and potash, sodium and calcium. Coal, particularly, may be the fuel of the future, and engineers are developing new methods of mining and processing it and of restoring the land after the coal has been extracted.

Starting salaries for mining engineers who go into the oil industry are $1,300 to $1,400 a month. Those starting in the mineral industries get about $1,000 to $1,200 a month. Petroleum engineers are the most sought after and can earn as much as $18,000 a year to start.

For more information, write:

Guidance Director, Mining,
Engineers' Council for Professional Development
345 East Forty-seventh Street
New York, New York 10017

BIOMEDICAL ENGINEERING

Because doctors lacked engineering experience and engineers lacked knowledge of the human body, a new specialty evolved called biomedical engineering. In order to provide doctors with the intricate machines needed to diagnose illnesses, and to measure and monitor the body's vital signs, engineers, working closely with doctors, began combining certain medical courses with their regular engineering courses, and created biomedical engineering.

If you choose this specialty, you will be working with doctors to save lives. You will be responsible for helping handicapped people lead more comfortable and easier lives. Or, you might design technology to help make the process of diagnosis more exact. Biomedical engineers have developed lasers that are used in removing cataracts, infrared rays for diagnosing cancer, computer-controlled artificial limbs, an artificial larynx, delicate hearing aids no larger than an earring, as well as such electronic miracles as pacemakers and electrocardiographs.

Just recently, a doctor-engineer team produced a manikin that can simulate the sounds and symptoms of fifty different heart conditions by the insertion of cassettes into its control panel. Dr. Michael Gordon, professor of medicine at the University of Miami, invented "Harvey," the manikin, which will be used to train medical students to recognize different types of heart disease. The electronics that make Harvey work were designed by Darrell G. Patterson, a former National Aeronautics and Space Administration (NASA) engineer who is now a research associate at the University of Miami. This is a perfect example of the engineer's skills turning the doctor's idea into reality.

Biomedical engineering is a fairly new field that has been growing since World War II. In 1972, there were three thousand biomedical engineers, and there will be openings for five thousand by 1985. There are probably more engineers in this field with advanced degrees than in most of the other specialties because so many are in research or teaching in universities.

Besides the regular engineering courses, such as calculus, chemistry, physics, computers, applied mechanics, and electric circuit theory, the biomedical engineering student might take animal biology, human physiology, clinical environments, biological modeling, and biomedical electronics. Many universities offer practical experience at a hospital and some arrange summer jobs for their students in this specialty.

For further information, write:

Engineers' Council for Professional Development
345 East Forty-seventh Street
New York, New York 10017

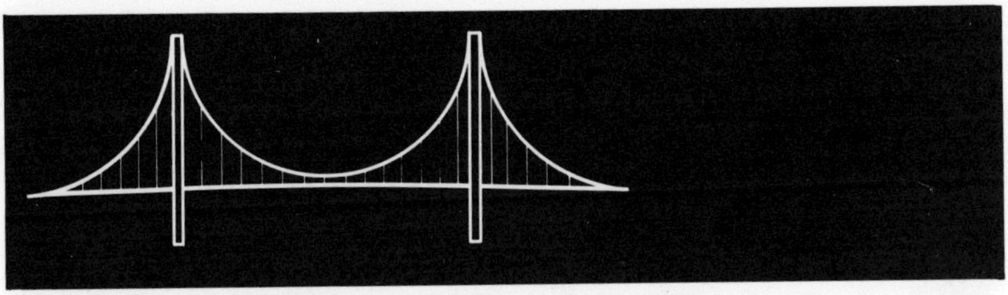

Careers in Engineering Technology

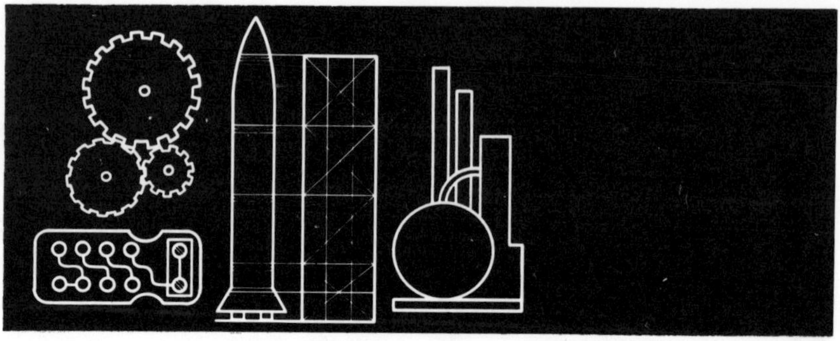

Engineering technology may be the career for you if you prefer the practical, "hands-on" application of engineering principles rather than the world of ideas and concepts. If you are the kind of person who is always taking things apart, have a knack for making things work, or a fascination for engines and motors, this might be the career for you. You would be working under the direction of an engineer. Even though you would be considered his or her assistant, you could eventually supervise others and do a great deal of innovative and challenging work yourself. Engineering technology is one of the few fields in which the technical assistant is encouraged to assume more and more responsibility so that the engineer can concentrate on design and development of new products and the creative end of production. In 1976, a higher percentage of technologists found jobs after graduation than engineers.

In line with the respect accorded the technologist, his or her starting salary today is about $12,500, far higher than assistants in other fields, and only about $2,000 less than that of the engineers. Of course, engineers earn much more money as they advance in their careers.

There is another type of technical assistant in engineering—the technician. It is important to distinguish between the technician and the technologist, because the technician earns a degree of Associate of Applied Science after two years of study in a junior or community college or technical institute, while the technologist studies for four years for his or her Bachelor of Technology degree. The technician may decide to stop after the two-year training period and take one of many jobs available assisting engineers or technologists. Technicians can also learn a great deal on the job and can increase the amount of individual responsibility they have. They may later elect to take the additional two years necessary to achieve a Bachelor of Technology degree. This degree can also be earned at night over a four-year period. Many companies will absorb part, or even all, of the tuition.

Technicians are much more limited in job opportunities than technologists, and their starting salary in 1977 was about $9,000 a year. That is still much higher than graduates of most other two-year programs.

The difference in training of the technologist and technician is mainly a difference in depth of knowledge. Obviously, the technician cannot learn in two years what the technologist took four years to learn. Technicians have less math and science than the technologists. Other subjects which they study include physics, communications, practical application of laboratory skills, construction methods, graphics, computation, testing, report writing, and equipment application and operation. The emphasis is on how to do things, rather than why they work the way they do, and technicians should share the engineer's and technologists enthusiasm for solving problems. Over 110 schools offer two-year associate degree programs, in such disciplines as metallurgical, electrical, mechanical, chemical, civil, and aeronautical engineering. Other specialties include air conditioning, heating and refrigeration, instrumentation, architectural and building construction, and tool design.

The technologist's degree is called either a Bachelor of Science in Engineering Technology or a Bachelor of Technology degree. The technologist requires basically the same

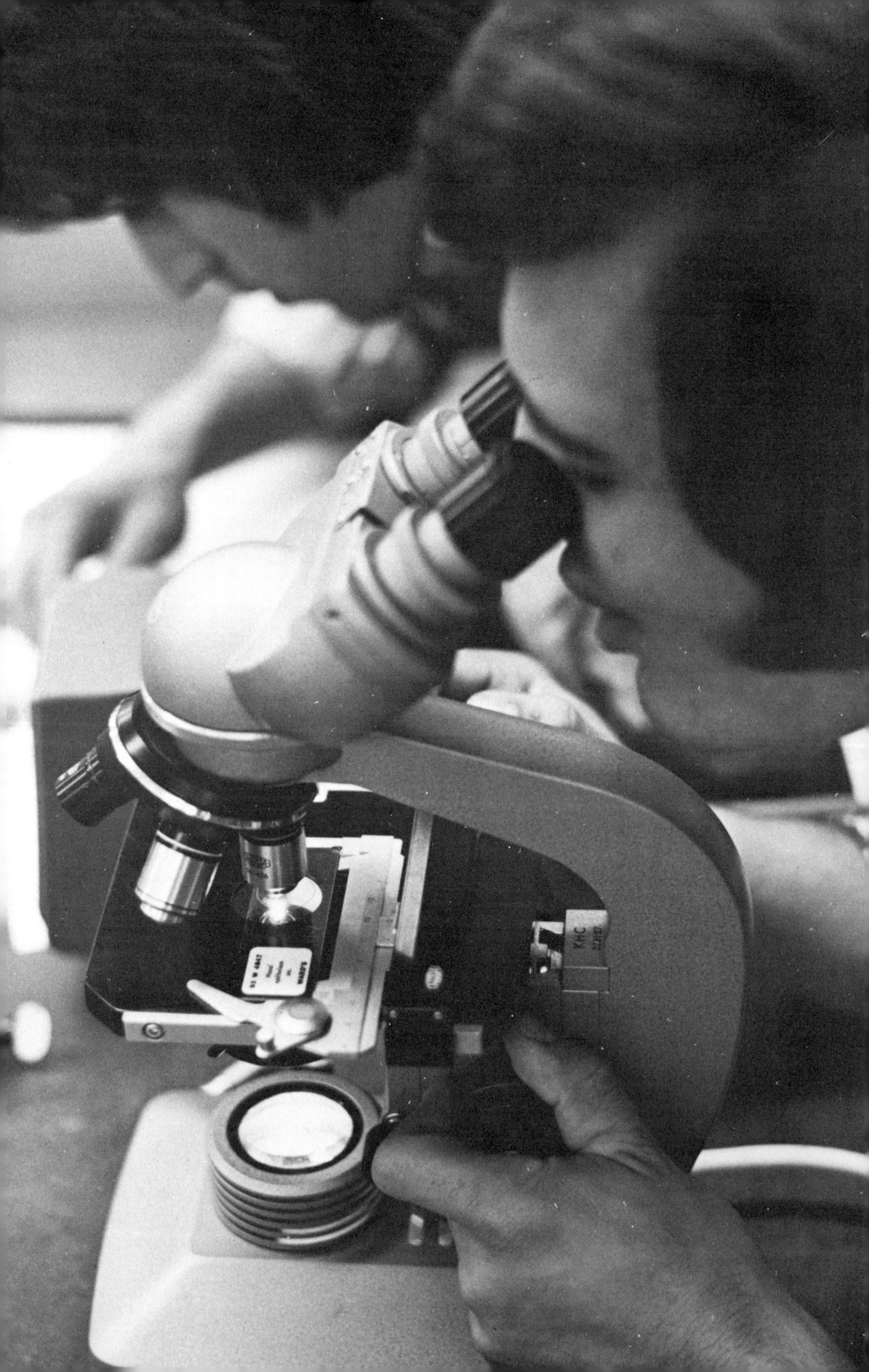

training as the technician, only takes more courses. Technologists study for four years (sometimes called two plus two), of which the first two may be taken in a two-year college, and the last two in a technological institute, or all four years in one school. Their curriculum includes more math and science than the technician takes, but less than the engineer. And technologists concentrate more on hardware (actual equipment) and processes than do engineers. Some schools continue in-depth training in the student's chosen specialty in the third and fourth years, and some, like the New York Institute of Technology, train students in a second technology to broaden their education. A student who specializes in electrical technology the first two years might learn about mechanical technology during the junior and senior years.

There were approximately one thousand graduates of the technical programs ten years ago, and today there are about sixty-five hundred a year, with room for more. The Department of Labor reports that Technologists will be marketable through 1985, no matter what area they train in.

Engineering technology is not a field most guidance counselors recommend to young women in high school. Only 12 percent of all technologists are women. However, if you are female, like math and science, and enjoy making things work, you should consider a career in engineering technology.

For more detailed information about careers as a technician or technologist, write to the Engineers' Council for Professional Development, 345 East Forty-seventh Street, New York, New York 10017.

To give you an idea of the specific duties of technicians and technologists in the major specialties, the chapters that follow will describe the experience of several of the people working in those fields.

A technician uses a high-powered microscope to do tests required by an engineer.

ELECTRICAL ENGINEERING TECHNICIANS AND TECHNOLOGISTS

Bob is a technician at Bell Laboratories. He earned his Associate of Applied Science degree at a community college and is now studying for his Bachelor of Electrical Technology degree at night, with his tuition paid for by Bell.

"I really enjoy my work—it's varied and interesting," Bob says. "I set up experiments in the lab according to my supervisor's instructions. Let's say there were some electron-microscope studies to be made of a crystalline, for example. The electrical engineer would give me a sample of the material to be studied and ask me to take pictures of it under the electron microscope."

As Bob gains more experience, he can work up to senior technical associate where he will have more responsibility and latitude to work on his own. An engineer would still be supervising his work, but Bob would exercise a great deal more individual judgment.

Electrical technicians and technologists are probably more in demand than those trained in any other area. Technicians and technologists trained in this field can find jobs in the broadcasting, communications, and automotive industries; in space-vehicle design, computer manufacturing, and medical electronics, to name a few possibilities. These people are needed in research and development, in production and maintenance, and in sales.

A high school background that includes algebra, trigonometry, physics, geometry, chemistry, English, shop, and lab work is the best preparation for this field.

The starting salary for technicians with an associate degree is about $9,000, and for technologists with a four-year Bachelor of Technology degree, about $12,500.

CHEMICAL ENGINEERING TECHNICIANS AND TECHNOLOGISTS

Linda is a valuable member of a production team at a large cosmetics factory. She is a senior technical aide to the chemi-

Edna Marshall, a switching technician, works with an engineer and a physicist at Bell Laboratories.

cal engineer in charge of her department. Linda graduated from a four-year program at a technological institute in Ohio where she earned a Bachelor of Science degree in chemical technology. She is thoroughly familiar with the equipment and procedures for chemical analyses, and supervises other technicians performing tests that were ordered by the engineer in charge.

"I've always loved chemistry, and I had a lot of encouragement from my dad," Linda says. "From the time I was a little child, he helped me in school and bought me chemistry sets and microscopes. Because I knew early that I wanted to go into chemistry, I went to a technical high school, where I took college chemistry courses and qualitative and quantitative analysis."

Lack of encouragement for girls interested in technical studies probably accounts for the low percentage of young women who do become technicians or technologists. However, this is a good time, male or female, to become one. John Schmid, director of placement at the New Jersey Institute of Technology, says: "There is a demand for chemical engineers and technologists now because of the need to produce new sources of energy and to cut down on air and water pollution. The supply is low, and the demand is high, so it's a good field to get into."

Salaries reflect the value of chemical technicians and technologists. The median salary for associate engineering technicians in this field in 1976 was $13,250. For chemical engineering technicians or technologists, it was $16,182. And for senior engineering technologists, it was $18,018. Experienced people were making more than $20,000.

Most chemical technicians and technologists work in chemical and petroleum plants, where they may be involved in production, development, quality control, research, or sales. They are also hired by dye, textile, perfume and food companies, and are sought by local, state and federal agencies concerned with pollution control and new energy sources. Jobs for those with an associate degree include doing routine tests; measuring heat, pressure and viscosity; and operating equipment. The more advanced technologists set up complicated experiments for the engineers and supervise technicians.

MECHANICAL ENGINEERING TECHNICIANS AND TECHNOLOGISTS

Drafting is the language of engineers, and in many companies, such as Ebasco Services in New York, the drafting is done by people like Larry, a technician trained in a two-year mechanical engineering technology program. He prepares the drawings for preliminary layouts as well as the final contract drawings. Larry's starting salary was $10,000 and as he gains more experience, he can make up to $20,000. "There's always a chance to learn and grow in this job," he says.

Mechanical technicians and technologists are needed in all manufacturing, power, communications, and construction industries in a wide variety of jobs. Some write sales brochures and manuals explaining how to service equipment or how to assemble it. Others are inspectors on production lines who monitor the quality of the item being manufactured or processed. In a bottling plant, for instance, a beginning technician might check to make sure the bottles hold the right amount of fluid. A more experienced technologist might be responsible for adjusting the machine that fills the bottles to make sure it doesn't overfill. Quality-control inspectors might also check screw threads to make sure the die that is used to manufacture them is still cutting the grooves deep enough.

A two- or four-year program in mechanical technology at a technical institute would include courses in drafting, characteristics of materials, gear systems, piping design, energy conversion and transmission systems, hydraulics, measurement instruments, welding, riveting, basic electricity, fluid power systems and theory, air conditioning, and heating systems and theory.

This is a good job for those persons who enjoy hands-on, practical application, who aren't afraid to get their hands dirty and jump right into the middle of things.

A student in the mechanical engineering technology laboratory at New York Institute of Technology.

Mechanical technology students at New York Institute of Technology converting gas powered automobile to electric power.

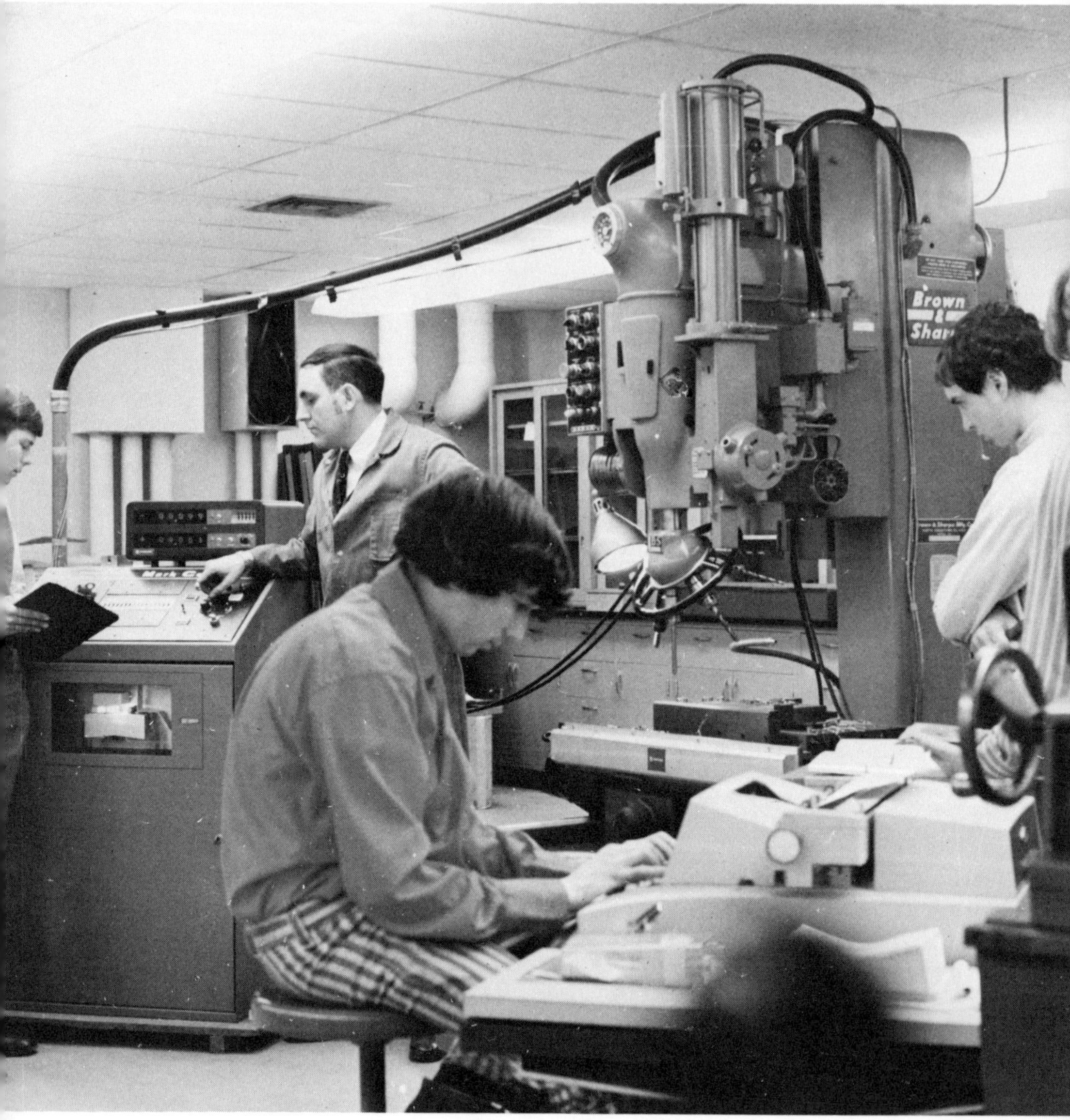

Mechanical technology students in a laboratory at New York Institute of Technology.

CIVIL ENGINEERING TECHNICIANS AND TECHNOLOGISTS

Jim works for a construction company as an assistant to a civil engineer. He took a two-year course at a community college where his training included learning concrete mixtures and quantities, inspection and materials testing, structural steel design, drafting, foundation design, piping design, soils analysis, cost estimating, construction equipment and functions, excavation and fill computations and planning, and surveying. He has an Associate of Applied Science degree and plans to earn a Bachelor of Technology degree at night.

"I enjoy working in the field," Jim says. "I like being outdoors. Part of my job is testing soils, testing concrete, foundation examinations. I'm also responsible for estimating how much material is needed on a job and how much it will cost. The engineer who supervises me is teaching me a lot about his job, and I hope to work more on my own after I have more experience."

An experienced civil engineering technologist might work on a highway project where he or she might be in charge of surveying, still under an engineer's supervision, but supervising those doing the surveying. A photogrammetrist, another kind of technologist, learns how to draw maps and charts based on aerial photographs. A specifications writer writes up the specifications for a construction job, with instructions from an engineer.

Civil engineers are needed to build bridges, highways, dams, housing projects, and skyscrapers; in power companies and in government agencies. They need technicians and technologists to help them, and many companies are increasing the number of technical aides to do the tasks that used to be assigned to beginning engineers. A summer job on a construction crew while you are in high school is a good way to get to know people who can advise you on the best training and experience to get in this field.

The median salary for technicians in the construction industry in 1976 was $13,775, with a top salary of $16,000. A supervisor can earn up to $18,500.

AIR CONDITIONING, HEATING, AND REFRIGERATION ENGINEERING TECHNICIANS AND TECHNOLOGISTS

Traditional methods of heating and cooling homes, factories, and office buildings will probably change drastically over the next twenty-five years. Because we are running out of fuels that we have always taken for granted, engineers must develop new equipment and new systems to conserve scarce fuels, and make use of new sources of energy. And wherever you find an engineer working to solve problems, you'll find technicians and technologists, too.

Bill is a salesman for an air conditioning company. He learned the basics of his field in a high school on-the-job training program, where he worked as a trainee in a shop that repaired air conditioning units. When one of the owners had a heart attack, he ran the shop and gained invaluable experience. He went into the service after high school and learned refrigeration technology. After the service, he worked in a place where he installed and repaired air-conditioners. He is now a salesman for his company and enjoys it, but misses the experience of actually working on the units.

Technicians and technologists who receive their training either at a technical institute or on the job, as Bill did, will be learning how to install, repair, improve, and test solar energy units and heat pumps. The new type of heating and cooling units being developed will revive this field in the future.

A technician with an associate degree might find a job with a design engineer testing and evaluating new units developed by the engineer, or might work in the research department of a company that tests heating and cooling units for spaceships.

Technologists with more training and experience might

An air-conditioning, heating, and refrigeration technician checks out an air-conditioner.

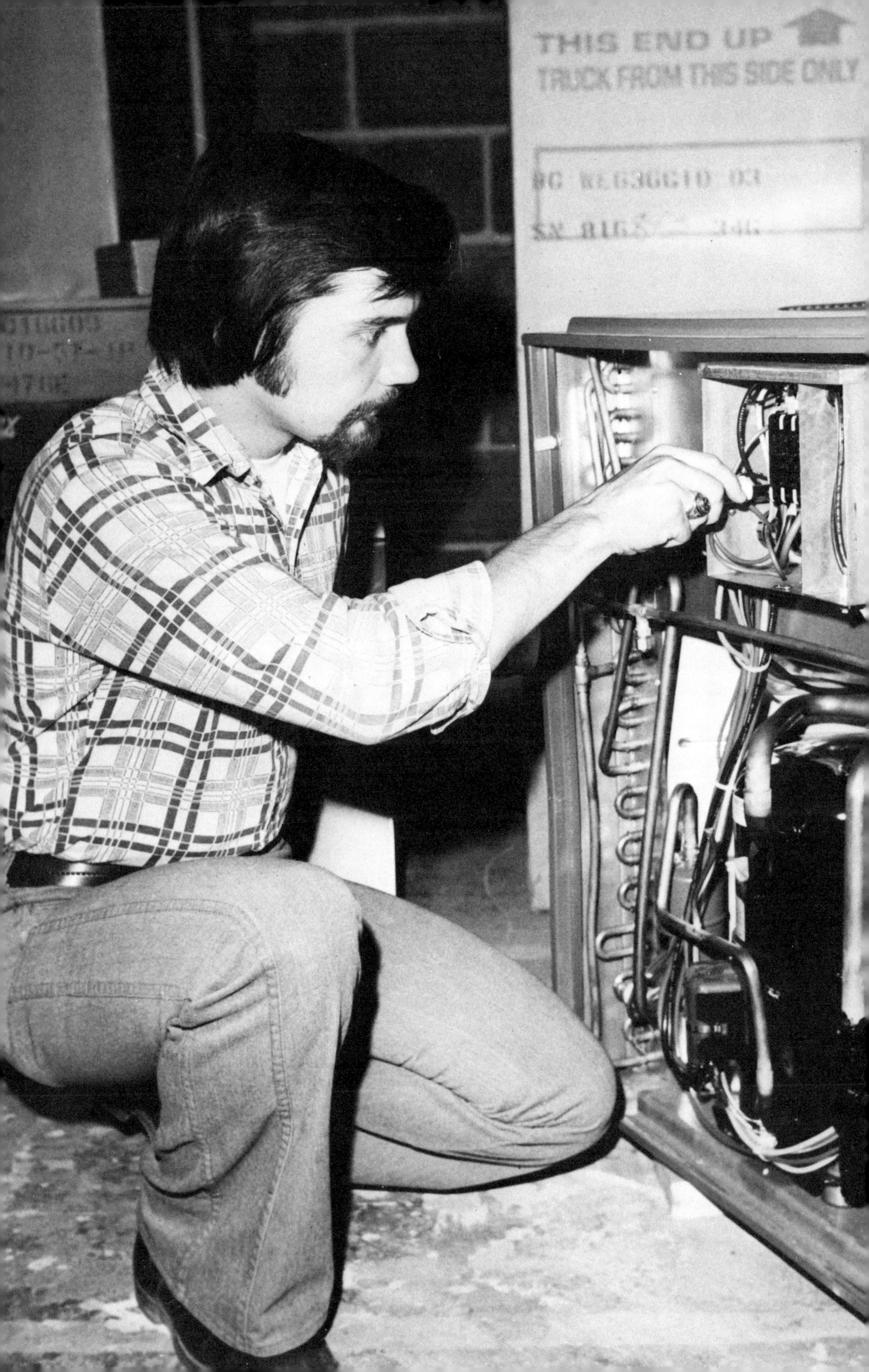
THIS END UP
TRUCK FROM THIS SIDE ONLY

make surveys and write proposals for a systems designer. They might travel around as manufacturers' representatives to show contractors how to install and maintain equipment. With enough experience, they might open their own air conditioning, heating, and refrigeration business.

Salaries start around $9,000 for maintenance people and raises are rapid as the technician's responsibilities increase.

AEROSPACE ENGINEERING TECHNICIANS AND TECHNOLOGISTS

Just as the aerospace engineer usually is trained in mechanical, electrical, civil, or some other branch of engineering and then specializes in aeronautics (aircraft) or astronautics (spacecraft), so the aerospace technician and technologist train in a specific field and then specialize in aerospace technology.

You will find the technician and technologist assisting the engineer in all areas of aerospace technology—research, design, development, manufacturing, sales, testing, and quality control. Their duties include maintenance of aircraft and spacecraft, inspection at all stages of manufacture, repairing equipment and systems, assembling models to be used in experiments, and testing various parts of the aircraft or spacecraft. They are also troubleshooters when a failure occurs.

Aerospace technicians and technologists are employed, not only by airplane manufacturers, but by the military and by helicopter and high-speed ground-transportation companies.

The median salary for technicians is $12,500, and for senior technical aides, $16,000. It is economical to hire them because there are many more jobs that can be done under the direction of engineers than there are jobs for engineers themselves.

Dean Herbert Fox of the New York Institute of Technology recently called an aerospace company to see if they were laying off technicians because of the cutback in defense funds. "They told me they were keeping the technicians and were laying off engineers," he said.

This field pays well and promises to pick up in the future as space exploration increases.

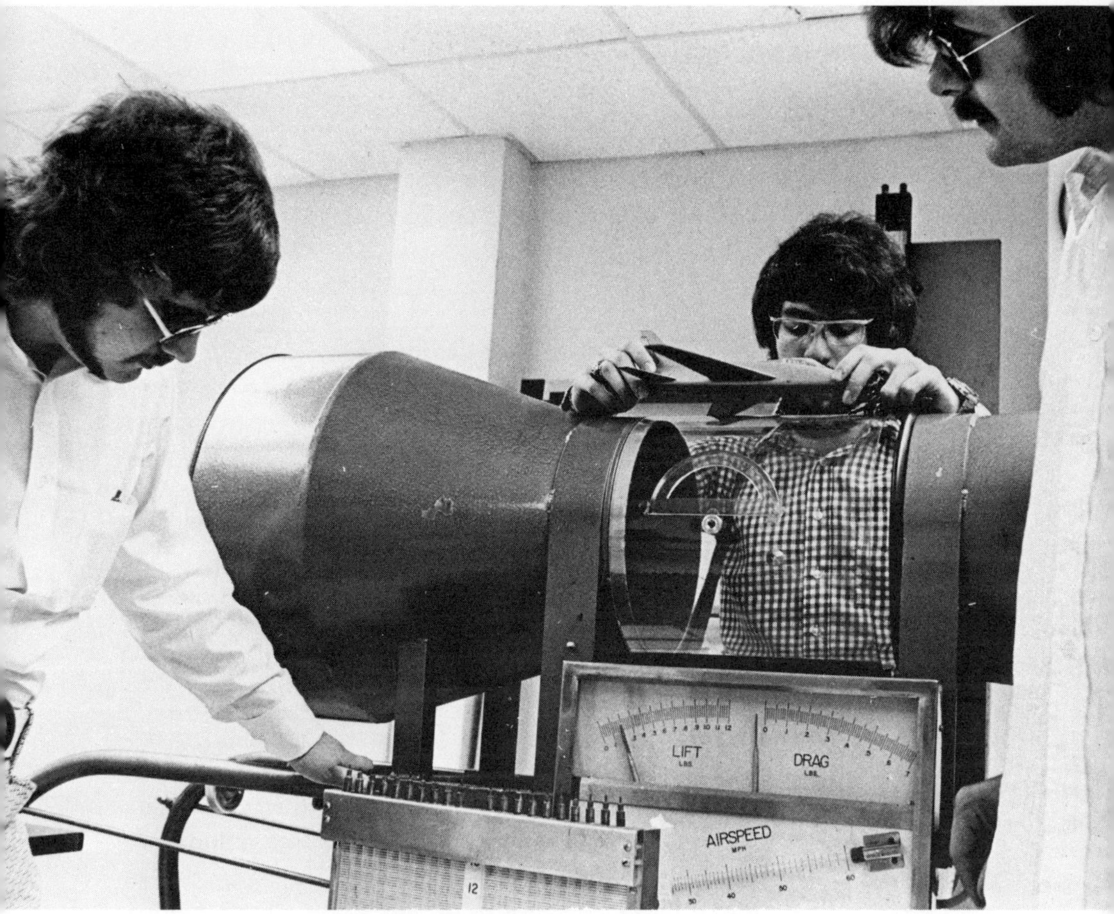

Aerospace technology students working on a wind tunnel in a lab at New York Institute of Technology.

INSTRUMENTATION TECHNICIANS AND TECHNOLOGISTS

As the instruments used in hospitals, industry, and laboratories become more and more complicated, a new type of technician is being trained to install, repair, test, maintain, and describe these instruments. Programs to teach people the intricacies of electronic, optical, hydraulic, pneumatic, or mechanical instruments are increasing in number in technical schools and in four-year colleges.

Ramapo College in New Jersey, for instance, awards a Bachelor of Science degree in instrumentation science after four years of study. The curriculum includes at least a year of basic sciences, including biology and chemistry, physics, calculus or algebra, geometry, and trigonometry. Later on in the program, students interested in biomedical engineering technology take organic chemistry, biochemistry, anatomy, and physiology. Other programs emphasize drafting, chemical engineering technology, or electronics technology.

The training in instrumentation for biomedical technology involves many kinds of complicated devices and machines. A technologist in this field knows how to attach an electrocardiograph or other monitoring machines to a patient. He or she knows how to install it, test it, and repair it if it fails. Technologists are trained to do routine tests on all hospital equipment to try to prevent failures before they happen, and to insure the safety of equipment used. Training in this speciality includes experience in every department of a hospital until all equipment is familiar to the student.

The median salary for biomedical technologists is $14,000. For supervisors, the median salary is $15,400.

In other fields, the instrumentation technician's duties include making sketches and diagrams of instruments—whether electrical or mechanical—according to the specifications of a design engineer. They make models, perform tests, and record evaluations for the engineer in charge.

Instrumentation science is a rapidly growing field with increasing need for technicians and technologists.

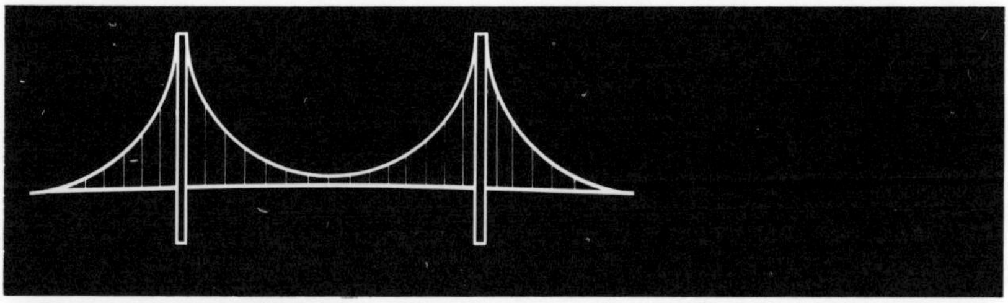

Appendix

Engineering Programs Accredited by the Engineers' Council for Professional Development

ACCREDITED PROGRAMS BY INSTITUTIONS

Air Force Institute of Technology
Wright-Patterson Air Force Base, OH 45433

Akron, University of
Akron, OH 44325

Alabama in Birmingham, University of
Birmingham, AL 35294

Alabama in Huntsville, University of
Huntsville, AL 35807

Alabama, The University of
University, AL 35486

Alaska, University of
Fairbanks, AK 99701

Alfred University
New York State College of Ceramics at Alfred, NY 14802

Arizona State University
Tempe, AZ 85281

Arizona, University of
Tucson, AZ 85721

Arkansas, University of
Fayetteville, AR 72701

Auburn University
Auburn, AL 36830

Boston University
Boston, MA 02215

Bradley University
Peoria, IL 61625

Bridgeport, University of
Bridgeport, CT 06602

Brigham Young University
Provo, UT 84602

Brown University
Providence, RI 02912

Bucknell University
Lewisburg, PA 17837

California Institute of Technology
Pasadena, CA 91125

California Polytechnic State University
San Luis Obispo, CA 93401

California State Polytechnic University, Pomona
Pomona, CA 91768

California State University, Chico
Chico, CA 95929

California State University, Fresno
Fresno, CA 93740

California State University, Fullerton
Fullerton, CA 92634

California State University, Long Beach
Long Beach, CA 90840

California State University, Los Angeles
Los Angeles, CA 90032

California State University, Northridge
Northridge, CA 91324

California State University, Sacramento
Sacramento, CA 95819

California, University of
Berkeley, CA 94720

California, University of
Davis, CA 95616

California, University of
Irvine, CA 92664

California, University of
Los Angeles, CA 90024

California, University of
Santa Barbara, CA 93106

Carnegie-Mellon University
Pittsburgh, PA 15213

Case Western Reserve University
Cleveland, OH 44106

Catholic University of America
Washington, DC 20017

Christian Brothers College
Memphis, TN 38104

Cincinnati, University of
Cincinnati, OH 45221

Citadel, The
Charleston, SC 29409

Clarkson College of Technology
Potsdam, NY 13676

Clemson University
Clemson, SC 29631

Cleveland State University
Cleveland, OH 44115

Colorado School of Mines
Golden, CO 80401

Colorado State University
Fort Collins, CO 80521

Colorado, University of
Boulder, CO 80302

Columbia University
New York, NY 10027

Connecticut, University of
Storrs, CT 06268

Cooper Union, The
New York, NY 10003

Cornell University
Ithaca, NY 14850

Dartmouth College
Thayer School of Engineering
Hanover, NH 03755

Dayton, University of
Dayton, OH 45469

Delaware, University of
Newark, DE 19711

Denver, University of
Denver, CO 80210

Detroit, University of
Detroit, MI 48221

Drexel University
Philadelphia, PA 19104

Duke University
Durham, NC 27706

Embry-Riddle Aeronautical University
Daytona Beach, FL 32014

Evansville, University of
Evansville, IN 47701

Fairleigh Dickinson University
Teaneck, NJ 07666

Federal City College
Washington, D.C. 20004

Florida Atlantic University
Boca Raton, FL 33431

Florida Institute of Technology
Melbourne, FL 32901

Florida Technological University
Orlando, FL 32816

Florida, University of
Gainesville, FL 32611

Gannon College
Erie, PA 16501

George Washington University
Washington, DC 20052

Georgia Institute of Technology
Atlanta, GA 30332

Georgia, University of
Athens, GA 30601

Hartford, University of
Hartford, CT 06117

Harvard University
Cambridge, MA 02138

Hawaii, University of
Honolulu, HI 96822

Hofstra University
Hempstead, NY 11550

Houston, University of
Houston, TX 77004

Howard University
Washington, DC 20001

Idaho, University of
Moscow, ID 83843

Illinois Institute of Technology
Chicago, IL 60616

Illinois at Chicago Circle, University of
Chicago, IL 60680

Illinois at Urbana—Champaign, University of
Urbana, IL 61801

Iowa State University
Ames, IA 50010

Iowa, University of
Iowa City, IA 52240

Johns Hopkins University
Baltimore, MD 21218

Kansas State University
Manhattan, KS 66506

Kansas, University of
Lawrence, KS 66045

Kentucky, University of
Lexington, KY 40506

Lafayette College
Easton, PA 18042

Lamar University
Beaumont, TX 77710

Lawrence Institute of Technology
Southfield, MI 48075

Lehigh University
Bethlehem, PA 18015

Louisiana State University
Baton Rouge, LA 70803

Louisiana Tech University
Ruston, LA 71270

Louisville, University of
Louisville, KY 40208

Lowell, University of
Lowell, MA 01854

Loyola Marymount University
Los Angeles, CA 90045

Maine at Orono, University of
Orono, ME 04473

Manhattan College
Riverdale, NY 10471

Marquette University
Milwaukee, WI 53233

Maryland, University of
College Park, MD 20742

Massachusetts Institute of Technology
Cambridge, MA 02139

Massachusetts, University of
Amherst, MA 01002

Memphis State University
Memphis, TN 38152

Merrimack College
North Andover, MA 01845

Miami, University of
Coral Gables, FL 33124

Michigan State University
East Lansing, MI 48823

Michigan Technological University
Houghton, MI 49931

Michigan, University of
Ann Arbor, MI 48104

Michigan, University of
Dearborn, MI 48128

Minnesota, University of
Minneapolis, MN 55455

Mississippi State University
Mississippi State, MS 39762

Mississippi, University of
University, MS 38677

Missouri—Columbia, University of
Columbia, MO 65201

Missouri-Rolla, University of
Rolla, MO 65401

Monmouth College
West Long Branch, NJ 07764

Montana College of Mineral Science and Technology
Butte, MT 59701

Montana State University
Bozeman, MT 59715

Naval Postgraduate School
Monterey, CA 93940

Nebraska, University of
Lincoln, NB 68588

Nevada-Reno, University of
Reno, NV 89507

New Hampshire, University of
Durham, NH 03824

New Haven, University of
West Haven, CT 06516

New Jersey Institute of Technology
Newark, NJ 07102

New Mexico Institute of Mining and Technology
Socorro, NM 87801

New Mexico State University
Las Cruces, NM 88001

New Mexico, University of
Albuquerque, NM 87131

New York, City College of the City University of
New York, NY 10031

New York at Buffalo, State University of
Buffalo, NY 14214

New York, State University of Maritime College
Ft. Schuyler, Bronx, NY 10465

New York at Stony Brook, State University of
Stony Brook, NY 11794

North Carolina Agricultural and Technical State University
Greensboro, NC 27411

North Carolina State University at Raleigh
Raleigh, NC 27607

North Carolina at Chapel Hill, University of
Chapel Hill, NC 27514

North Carolina at Charlotte, University of
Charlotte, NC 28223

North Dakota State University
Fargo, ND 58102

North Dakota, University of
Grand Forks, ND 58201

Northeastern University
Boston, MA 02115

Northern Arizona University
Flagstaff, AZ 86001

Northrop University
Inglewood, CA 90306

Northwestern University
Evanston, IL 60201

Norwich University
Northfield, VT 05663

Notre Dame, University of
Notre Dame, IN 46556

Oakland University
Rochester, MI 48063

Ohio Northern University
Ada, OH 45810

Ohio State University
Columbus, OH 43210

Ohio University
Athens, OH 45701

Oklahoma State University
Stillwater, OK 74074

Oklahoma, University of
Norman, OK 73069

Old Dominion University
Norfolk, VA 23508

Oregon State University
Corvallis, OR 97331

Pacific, University of the
Stockton, CA 95211

Pennsylvania State University, The
University Park, PA 16802

Pennsylvania, University of
Philadelphia, PA 19104

Pittsburgh, University of
Pittsburgh, PA 15213

Polytechnic Institute of New York
Brooklyn, NY 11201

Portland State University
Portland, OR 97207

Portland, University of
Portland, OR 97203

Prairie View A & M University
Prairie View, TX 77445

Pratt Institute
Brooklyn, NY 11205

Princeton University
Princeton, NJ 08540

Puerto Rico, University of
Mayaguez, PR 00708

Purdue University
W. Lafayette, IN 47907

Rensselaer Polytechnic Institute
Troy, NY 12181

Rhode Island, University of
Kingston, RI 02881

Rice University
Houston, TX 77001

Rochester Institute of Technology
Rochester, NY 14623

Rochester, University of
Rochester, NY 14627

Rose-Hulman Institute of Technology
Terre Haute, IN 47803

Rutgers University—The State University of New Jersey
New Brunswick, NJ 08903

St. Martin's College
Olympia, WA 98503

San Diego State University
San Diego, CA 92182

San Francisco State University
San Francisco, CA 94132

San Jose State University
San Jose, CA 95192

Santa Clara, University of
Santa Clara, CA 95053

Seattle University
Seattle, WA 98122

South Carolina, University of
Columbia, SC 29208

South Dakota School of Mines and Technology
Rapid City, SD 57701

South Dakota State University
Brookings, SD 57006

South Florida, University of
Tampa, FL 33602

Southeastern Massachusetts University
North Dartmouth, MA 02747

Southern California, University of
Los Angeles, CA 90007

Southern Illinois University at Carbondale
Carbondale, IL 62901

Southern Illinois University at Edwardsville
Edwardsville, IL 62025

Southern Methodist University
Dallas, TX 75275

Southern University
Baton Rouge, LA 70813

Southwestern Louisiana, University of
Lafayette, LA 70501

Stanford University
Stanford, CA 94305

Stevens Institute of Technology
Hoboken, NJ 07030

Swarthmore College
Swarthmore, PA 19081

Syracuse University
Syracuse, NY 13210

Tennessee State University
Nashville, TN 37203

Tennessee Technological University
Cookeville, TN 38501

Tennessee, University of
Knoxville, TN 37916

Texas A & I University
Kingsville, TX 78363

Texas A & M University
College Station, TX 77843

Texas Tech University
Lubbock, TX 79409

Texas at Arlington, University of
Arlington, TX 76010

Texas at Austin, University of
Austin, TX 78712

Texas at El Paso, University of
El Paso, TX 79968

Toledo, University of
Toledo, OH 43606

Trinity University
San Antonio, TX 78284

Tri-State University
Angola, IN 46703

Tufts University
Medford, MA 02155

Tulane University
New Orleans, LA 70118

Tulsa, University of
Tulsa, OK 74104

Tuskegee Institute
Tuskegee Institute, AL 36088

Union College
Schenectady, NY 12308

United States Air Force Academy
USAF Academy, CO 80840

United States Coast Guard Academy
New London, CT 06320

United States Naval Academy
Annapolis, MD 21402

Utah State University
Logan, UT 84322

Utah, University of
Salt Lake City, UT 84112

Valparaiso University
Valparaiso, IN 46383

Vanderbilt University
Nashville, TN 37203

Vermont, University of
Burlington, VT 05401

Villanova University
Villanova, PA 19085

Virginia Military Institute
Lexington, VA 24450

Virginia Polytechnic Institute and State University
Blacksburg, VA 24061

Virginia, University of
Charlottesville, VA 22901

Walla Walla College
College Place, WA 99324

Washington State University
Pullman, WA 99163

Washington, University of
Seattle, WA 98195

Washington University
St. Louis, MO 63130

Wayne State University
Detroit, MI 48202

Webb Institute of Naval Architecture
Glen Cove, NY 11542

West Virginia Institute of Technology
Montgomery, WV 25136

West Virginia University
Morgantown, WV 26506

Western New England College
Springfield, MA 01119

Wichita State University
Wichita, KS 67208

Widener College
Chester, PA 19013

Wisconsin—Madison, University of
Madison, WI 53706

Wisconsin—Milwaukee, University of
Milwaukee, WI 53201

Wisconsin—Platteville, University of
Platteville, WI 53818

Worcester Polytechnic Institute
Worcester, MA 01609

Wyoming, University of
Laramie, WY 82071

Yale University
New Haven, CT 06520

Youngstown State University
Youngstown, OH 44503

Engineering Technology Programs Accredited by the Engineers' Council for Professional Development

ACCREDITED PROGRAMS BY INSTITUTION:

Academy of Aeronautics
LaGuardia Airport
Flushing, NY 11371

Akron, University of
Community & Technical College
Akron, OH 44325

Alamance, Technical Institute of
Burlington, NC 27215

Anoka-Ramsey Community College
Coon Rapids, MN 55433

Arizona State University
Tempe, AZ 85281

Belleville Area College
Belleville, IL 62221

Blue Mountain Community College
Pendleton, OR 97801

Bluefield State College
Bluefield, WV 24701

Bradley University
Peoria, IL 61625

Brigham Young University
Provo, UT 84602

Bronx Community College
Bronx, NY 10453

Broome Community College
Binghamton, NY 13902

California Polytechnic State University
San Luis Obispo, CA 93407

California State Polytechnic University
Pomona, CA 91768

Capitol Institute of Technology
Kensington, MD 20795

Chattanooga State Technical Community College
Chattanooga, TN 37406

Cincinnati, University of
Ohio College of Applied Science
Cincinnati, OH 45210

Cogswell College
San Francisco, CA 94108

Colorado Technical College
Manitou Springs, CO 80829

Columbus Technical Institute
Columbus, OH 43215

Community & Technical College
A School of West Virginia Institute of Technology
Montgomery, WV 25136

Dayton, University of
Dayton, OH 45469

Del Mar College
Corpus Christi, TX 78404

Delta College
University Center, MI 48710

DeVry Institute of Technology
Atlanta, GA 30308

DeVry Institute of Technology
Chicago, IL 60618

DeVry Institute of Technology
Dallas, TX 75235

DeVry Institute of Technology
Phoenix, AZ 85016

District of Columbia, University of
the Van Ness Campus
Washington, DC 20008

Eastern New Mexico University Technical Institute
Portales, NM 88130

Embry-Riddle Aeronautical University
Daytona Beach, FL 32015

Erie Community College North Campus
Buffalo, NY 14221

Fayetteville Technical Institute
Fayetteville, NC 28303

Florence-Darlington Technical College
Florence, SC 29501

Florida A & M University
Tallahassee, FL 32307

Forsyth Technical Institute
Winston-Salem, NC 27103

Franklin Institute of Boston
Boston, MA 02116

Gaston College
Dallas, NC 28034

Grossmont College
El Cajon, CA 92020

Guilford Technical Institute
Jamestown, NC 27282

Hartford State Technical College
Hartford, CT 06106

Hartford, University of Samuel I. Ward Technical College
West Hartford, CT 06117

Hawkeye Institute of Technology
Wateroo, IA 50704

Houston, University of College of Technology
Houston, TX 77004

Hudson Valley Community College
Troy, NY 12180

Indiana University—Purdue University at Fort Wayne
Fort Wayne, IN 46805

Indiana University—Purdue University at Indianapolis
Indianapois, IN 46205

Kansas Technical Institute
Salina, KS 67401

Kent State University Tuscarawas Campus
New Philadelphia, OH 44663

Lake Superior State College
Sault Ste. Marie, MI 49783

Lowell, University of
Lowell, MA 01854

Maine at Orono, University of
Orono, ME 04473

Memphis State University
Memphis, TN 38152

Memphis, State Technical Institute at
Memphis, TN 38134

Mercer County Community College
Trenton, NJ 08690

Michigan Technological University
Houghton, MI 49931

Middlesex County College
Edison, NJ 08817

Midlands Technical College
Columbia, SC 29250

Milwaukee School of Engineering
Milwaukee, WI 53201

Missouri Institute of Technology
Kansas City, MO 64108

Mohawk Valley Community College
Utica, NY 13501

Montana State University
Bozeman, MT 59715

Morrison Institute of Technology
Morrison, IL 61270

Nashville State Technical Institute
Nashville, TN 37209

Nassau Community College
Garden City, NY 11530

Nebraska at Omaha, University of
Omaha, NB 68101

Nevada, University of
College of Engineering
Reno, NV 89507

New Hampshire Technical Institute
Concord, NH 03301

New Mexico State University
Las Cruces, NM 88003

New York Institute of Technology
Metropolitan Campus
New York, NY 10001

New York Institute of Technology
Old Westbury, NY 11568

New York, State University of,
Agricultural & Technical College
Alfred, NY 14802

New York, State University of,
Agricultural & Technical College
Canton, NY 13617

New York, State University of,
Agricultural & Technical College
Farmingdale, NY 11735

North Carolina at Charlotte,
University of
Charlotte, NC 28223

Northeastern University
Lincoln College
Boston, MA 02115

Northern Arizona University
Flagstaff, AZ 86001

Northrop University
Inglewood, CA 90306

Norwalk State Technical College
Norwalk, CT 06854

Ocean County College
Toms River, NJ 08753

Ohio Institute of Technology
Columbus, OH 43209

Oklahoma State University
School of Technology
Stillwater, OK 74074

Old Dominion University
Norfolk, VA 23508

Oregon Institute of Technology
Klamath Falls, OR 97601

Oregon State University
Corvallis, OR 97331

Parkland College
Champaign, IL 61820

Pennsylvania State University
Altoona Campus
Altoona, PA 16603

Pennsylvania State University
Beaver Campus
Monaca, PA 15061

Pennsylvania State University
Behrend College
Erie, PA 16510

Pennsylvania State University
Berks Campus
Reading, PA 19608

Pennsylvania State University
Capitol Campus
Middletown, PA 17057

Pennsylvania State University
Delaware County Campus
Media, PA 19063

Pennsylvania State University
DuBois Campus
DuBois, PA 15801

Pennsylvania State University
Fayette Campus
Uniontown, PA 15401

Pennsylvania State University
Hazleton Campus
Hazleton, PA 18201

Pennsylvania State University
McKeesport Campus
McKeesport, PA 15132

Pennsylvania State University
Mont Alto Campus
Mont Alto, PA 17237

Pennsylvania State University
New Kensington Campus
New Kensington, PA 15068

Pennsylvania State University
Ogontz Campus
Abington, PA 19001

Pennsylvania State University
Schuylkill Campus
Schuylkill Haven, PA 17972

Pennsylvania State University
Shenango Valley Campus
Sharon, PA 16146

Pennsylvania State University
Wilkes-Barre Campus
Wilkes-Barre, PA 18708

Pennsylvania State University
Worthington Scranton Campus
Dunmore, PA 18512

Pennsylvania State University
York Campus
York, PA 17403

Phoenix College
Phoenix, AZ 85013

Piedmont Technical College
Greenwood, SC 29646

Pittsburgh at Johnstown,
University of
Johnstown, PA 15904

Purdue University Calumet Campus
Hammond, IN 46323

Purdue University
West Lafayette Campus
West Lafayette, IN 47907

Queensborough Community College
of the City University of New York
Bayside, NY 11364

Ricks College
Rexburg, ID 83440

Rochester Community College
Rochester, MN 55901

Rochester Institute of Technology
Rochester, NY 14623

St. Louis Community College
at Florissant Valley
St. Louis, MO 63135

St. Petersburg Junior College
St. Petersburg, FL 33733

San Francisco, City College of
San Francisco, CA 94112

Savannah State College
Savannah, GA 31404

Sinclair Community College
Dayton, OH 45402

Southeastern Massachusetts
University
North Dartmouth, MA 02747

Southern Colorado, University of
Pueblo, CO 81001

Southern Illinois University
at Carbondale
Carbondale, IL 62901

Southern Technical Institute
A Division of Georgia Institute
of Technology
Marietta, GA 30060

Spartanburg Technical College
Spartanburg, SC 29303

Spring Garden College
Chestnut Hill, PA 19118

Sumter Area Technical College
Sumter, SC 29150

Temple University
College of Engineering Technology
Philadelphia, PA 19122

Tennessee at Martin, University of
Martin, TN 38238

Texas A & M University
College Station, TX 77840

Texas Tech University
Lubbock, TX 79409

Thames Valley State Technical College
Norwich, CT 06360

Toledo, University of
Community & Technical College
Toledo, OH 43606

Vermont Technical College
Randolph Center, VT 05061

Virginia Polytechnic Institute & State University
Blacksburg, VA 24061

Wake Technical Institute
Raleigh, NC 27603

Waterbury State Technical College
Waterbury, CT 06708

Weber State College
Ogden, UT 84408

Wentworth Institute
Boston, MA 02115

Western Kentucky University
Bowling Green, KY 42101

Youngstown State University
Youngstown, OH 44503

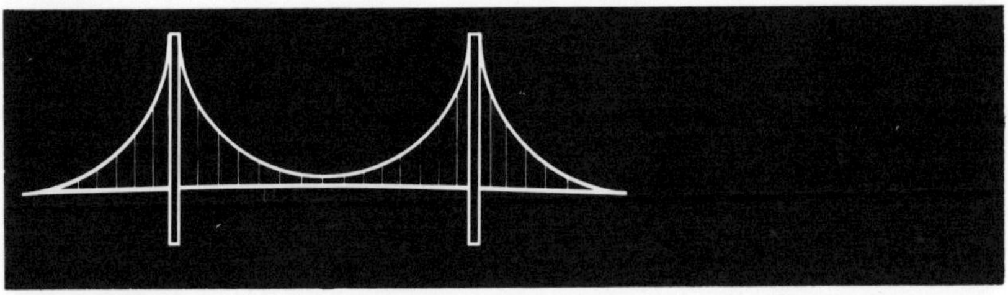

Index